Ralph Waldo Trine

Every Living Creature

Heart-Training Through the Animal World

Ralph Waldo Trine

Every Living Creature
Heart-Training Through the Animal World

ISBN/EAN: 9783744734240

Printed in Europe, USA, Canada, Australia, Japan

Cover: Foto ©berggeist007 / pixelio.de

More available books at **www.hansebooks.com**

Every Living Creature

OR HEART-TRAINING THROUGH THE ANIMAL WORLD

By

RALPH WALDO TRINE

The tender and humane passion in the human heart is too precious a quality to allow it to be hardened or effaced by practices such as we so often indulge in

———

NEW YORK

THOMAS Y. CROWELL & CO.

PUBLISHERS

EVERY LIVING CREATURE

OR

HEART-TRAINING THROUGH THE ANIMAL WORLD

IT is said that in Japan if one picks up a stone to throw at a dog, the dog will not run, as you will find he will in almost every case here, because *there* the dog has never had a stone thrown at him, and consequently he does not know what it means. This spirit of gentleness, kindliness, and care for the animal world is a characteristic of the Japanese people. It in turn manifests itself in all of their relations with their fellow-men; and one of the results is that the amount of crime committed there each year in proportion to the population is but a very small fraction of that committed in the United States.

In India, where the treatment of the animal world is something to put to shame

A

our own country, with its boasted Christian civilisation and power, there, with a population of some three hundred millions, there is but one-fourth the amount of crime that there is each year in England, with a population of some twenty millions, and only a fraction of what it is in the United States, with a population of not more than one-fourth the population of India. These are most significant facts; they are indeed facts of tremendous import, and we would do wisely to estimate them at their proper value.

We cannot begin too early in inculcating what I would term humane sentiments in the mind and heart of every individual. How early and almost unconsciously the mother, for example, gives the first lessons of thoughtlessness, carelessness, and what will eventually result in cruelty or even crime, to her child. The child is put upon the hobby-horse, a whip is put into his little hand, and he is told: "Now whip the old horse and make him go." With this initial lesson, continued in various ways, we find the eager desire the child has for whipping, when he gets the whip into his hands in a waggon behind a real horse. Or even when younger, the child stumbles over a chair, receives a knock, and bursts into crying. The mother, in some cases merely thought-less, in others caring only for her own comfort and ease, in order to call the atten-tion of the child away from the little hurt and greater rage and fright, says: "Did the mean chair hurt mamma's little boy? Go and kick the old chair—kick it hard." The next day when the child falls over or bumps against the dog, the dog in turn is the one to receive the kick; and still later, when anything of the kind occurs in connection with a little playmate, the playmate re-ceives the same treatment. And, so far as his relations with his fellow-men, when he is grown to manhood, are concerned, each one can trace them for himself.

We have sketched the thoughtless or the selfish mother. Let us look for a moment at the other type of mother, the one who is ever thoughtful, desirous of bringing the best influence to bear upon this little sensi-tive plate, if you will allow the expression, the mother who understands the great, almost omnipotent forming-power of early impres-sions. The child stumbles over or falls against the chair. The mother, after smoothing the hurt place and kissing away the first impulse to anger and also the fright of the child, and thereby its tears, says: "And now I wonder if mamma's little boy has hurt the chair. Go bring it to mamma and let her smooth away its hurt also." This is done, and all is now as if nothing had occurred. The next day, then, when the child stumbles over or bumps against the dog, after he has had his own hurt soothed by his mother, he in turn toddles off to soothe and comfort the dog; and again, when the child bumps against his little play-fellow, after he has been soothed and kissed and thereby comforted by his mother, he feels for and sympathises with the other little fellow, and brings him up to receive the same treatment. And again, each one can for himself carry the effects of this type of suggestion and training into the child's later life and into his relations with his fellow-men. Many instances of this nature in the every-day life of the mother and child might be mentioned.

And to go back even farther—those mothers who are beginning to understand the powerful moulding influences of pre-natal conditions will realise that every mental and emotional state lived in by the mother makes its influence felt in the life of the forming child, and she will therefore be careful that during the period she is carrying the child no thoughts or emotions of anger, or hatred, or envy, or malice, no unkind thoughts of any kind be entertained by her, but, on the contrary, thoughts of tenderness, kindness, compassion, and love ; these then will influence and lead the mind of the child when born, and will in turn externalise their effects in his body, instead of allowing to be externalised the poisoning and destructive effects of their opposites.

HEART-TRAINING

It is an established fact that the training of the intellect alone is not sufficient. Nothing in this world can be truer than that the education of the head, without the training of the heart, simply increases one's power for evil, while the education of the heart, along with the head, increases one's power for good, and this, indeed, is the true education.

Clearly we must begin with the child. The lessons learned in childhood are the last to be forgotten. The potter moulds the clay only when it is soft; in a little while, when it begins to harden, he has no more power over it. So it is with the child. The first principles of conduct instilled into his mind, planted within his heart, take root and grow, and as he grows from childhood to youth, and from youth to manhood, these principles become fixed. They exert their influence. Scarcely any power in existence can change them. They cling to him through life. They decide his destiny. How important, then, that these first principles implanted within the child's heart be lessons of gentleness, kindness, mercy, love, and humanity, and not lessons of hatred, envy, selfishness, and malice! The former make ultimately our esteemed, law-abiding, law-loving citizens; the latter law-breakers and criminals. Upon the training of the children of to-day depends the condition of our country a generation hence.

In crimes against the person the passions play the most important part, and this is true, also, even in many crimes against property. How important it is, then, that the child be taught to govern its passions! How important that it be taught to be kind, gentle, loving, and humane; and in all the range of human thought there is not a better, wiser, or more expedient way of accomplishing this end than by teaching kindness towards God's lower creatures. If children are thus taught they will have instilled into their hearts those principles of action which will make them kind and merciful not only to the lower animals, but also toward their fellow-men as they attain to manhood. Let them be taught that the lower animals are God's creatures, as they themselves are, put here by a common Heavenly Father, each for its own special purpose, *and that they have the same right to life and protection.* Let them be taught that principle recognised by all noble - hearted men, that it is only a depraved, debased, and cowardly nature that will injure an inferior, defenceless creature, simply because it is in its power to do so, and that there is no better, no grander test of true bravery and nobility of character than one's treatment of the lower animals.

It is impossible to over-estimate the benefits resulting from judicious, humane instruction. The child who has been taught nothing of mercy, nothing of humanity, who has never

been brought to realise the claims that
animals have upon him for protection and
kindness, will grow up to be thoughtless and
cruel toward them, and if he is cruel to them
that same heart, untouched by kindness and
mercy, will prompt him to be cruel to his
family, to his fellow - men. On the other
hand, the child who has been taught to
realise the claims that God's lower creatures
have upon him, whose heart has been touched
by lessons of kindness and mercy, under their
sweet influence will grow to be a large-
hearted, tender-hearted, manly man. Then
let the children be trained, their hands, their
intellects, and above all their hearts. Let
them be taught to have pity for the animals
that are at our mercy, that cannot protect
themselves, that cannot explain their weakness, their pain, or their suffering, and soon
this will bring to their recognition that higher
law, the moral obligation of man as a superior
being to protect and care for the weak and
defenceless. Nor will it stop here, for this
in turn will lead them to that highest law—
man's duty to man.

Hunting

So great do I believe are the influences of the inculcation of humane sentiments early in the life of every individual that I shall endeavour to make as concrete as possible the suggestions which are to follow; for criminal training or humane training can be and is continually given in numbers of ways.

As a parent, in the first place, I would teach the child the thoughtlessness, the selfishness, the heartlessness, the cruelty of hunting for sport. I would put into his hands no air-guns or instruments or weapons by which he could inflict torture upon or take the life of birds or other animals. Instead of encouraging him in torturing or killing the birds, I would point out to him the great service they are continually doing for us in the destruction of various worms and insects and small rodents which, if left to themselves, would so multiply as literally to destroy practically all fruit and plant life. I would have him remember how many lives are enriched and beautified by their song. I would point out to him their habits of industry, their

marvellous powers of adaptation, their insight and perseverance. Therefore I would
teach him to love, to study, to care for
and feed them.

Hunting for sport indicates one of two
things — a nature of such thoughtlessness
as to be almost inexcusable, or a selfishness
so deplorable as to be unworthy a normal,
sane human being. *No truly thoughtful
manly man or truly thoughtful womanly
woman will engage in it.* And when we
read of this or that woman, be she well
known in society, or the wife of this or
that well-known man, so following her
selfish, savage, cruel instincts, or her desire for notoriety or newspaper comments,
as to take part in a deer-hunt, a fox-chase,
or in a hunt of any type, we have an index
to her real character that should be sufficient.

But a few days ago my attention was
called to a minister in one of the New
England cities, who had come out in the
papers with an article on hunting as a most
excellent pastime and recreation for the
members of his calling, and urged them
to take it up, as he already had. Think
of it, what it means, — a man who has

gotten no farther into the real spirit of the gentle and compassionate teachings of the Christ whom he professes to follow, to say nothing of the humane teachings of the gentle Buddha, whom this reverend gentleman would, by the way, refer to in his pulpit and his prayer-meetings as the heathen! Shall we refrain from saying, inexcusable thoughtlessness, or brutal, deplorable selfishness? I cannot refrain in this connection from quoting a sentence or two from Archdeacon Farrar which have recently come to my notice:

"Not once or twice only, at the seaside, have I come across a sad and disgraceful sight—a sight which haunts me still—a number of harmless sea-birds lying defaced and dead upon the sand, their white plumage red with blood, as they had been tossed there, dead or half-dead, their torture and massacre having furnished a day's amusement to heartless and senseless men. Amusement! I say execrable amusement! All killing for mere killing's sake is execrable amusement. Can you imagine the stupid callousness, the utter insensibility to mercy and beauty, of the man who, seeing those bright, beautiful creatures as their white, immaculate wings flash in the sunshine over the blue waves, can go out in a boat with his boys to teach them to become brutes in character by finding amusement—I say, again, dis-humanising amusement — by wantonly murdering these fair birds of God, or cruelly wounding them, and letting them fly away to wait and die in lonely places?"

And another paragraph which was sent me by a kind friend to our fellow-creatures a few days ago:

"The celebrated Russian novelist, Turgenieff, tells a most touching incident from his own life, which awakened in him sentiments that have coloured all his writings with a deep and tender feeling.

"When Turgenieff was a boy of ten his father took him out one day bird-shooting. As they tramped across the brown stubble, a golden pheasant rose with a low whirr from the ground at his feet, and, with the joy of a sportsman throbbing through his veins, he raised his gun and fired, wild with excitement when the creature fell fluttering at his side. Life was ebbing fast, but the instinct of the mother was stronger than death itself, and with a feeble flutter of her wings the mother bird reached the nest where her young brood were huddled, unconscious of danger. Then, with such a look of pleading and reproach that his heart stood still at the ruin he had wrought,— and never to his dying day did he forget the feeling of cruelty and guilt that came to him in that moment,—the little brown head toppled over, and only the dead body of the mother shielded her nestlings.

"'Father, father,' he cried, 'what have I done?' as he turned his horror-stricken face to his father. But not to his father's eye had this little tragedy been enacted, and he said: 'Well done, my son; that was well done for your first shot. You will soon be a fine sportsman.'

"'Never, father; never again shall I destroy any living creature. If that is sport I will have none of it. Life is more beautiful to me than death, and since I cannot give life, I will not take it.'"

And so, instead of putting into the hands of the child a gun or any other weapon that may be instrumental in crippling, torturing, or taking the life of even a

single animal, I would give him the field-
glass and the camera, and send him out
to be a friend to the animals, to observe
and study their characteristics, their habits,
to learn from them those wonderful lessons
that can be learned, and thus have his
whole nature expand in admiration and
love and care for them, and become
thereby the truly manly and princely type
of man, rather than the careless, callous,
brutal type.

VIVISECTION

Another practice let us consider that is
clearly hardening in its influence—a practice that children and older students are
here and there called upon to witness. I
refer to the practice commonly known as
vivisection—the cutting, freezing, burning,
tearing, torturing of live animals for purposes of scientific "investigation."   After
making a most careful study of this matter
and its claims, getting the opinions of
many of the ablest physicians and surgeons
in the world, I have been forced to come
to the conclusion that practically nothing of
any *real* value has come to us through this

channel that could not and would not have come in other ways without this great torture and sacrifice of life, to say nothing of the cruel and hardening effects upon those who resort to these methods.

Personally, I should allow no child of mine to attend or remain at any school where it is carried on, and, moreover, I should raise my voice and exert my influence against it at every opportunity. I should teach the child the great fact that we are so rapidly learning to-day—namely, that the mind is the natural protector of the body, and that there are being continually externalised in the body, effects and conditions most akin to our prevailing mental states and emotions. I should teach him that it is unwise as well as cowardly to bring diseased conditions into the body through the poisoning, corroding effects of anger, hatred, jealousy, malice, envy, rage, fear, worry, lust, intemperance, and then seek to find an aid to the remedy through the torture of even a single dumb fellow-creature.

DOCKING

In the next place, as an object-lesson, I should point out to the child what is indicated at the sight of a dock-tailed horse. It indicates one of two things—weakness of individuality and hence slavery to custom, or that all too-prevalent vain desire through parade to attract attention, because the owner of the animal is conscious of the fact that there is not enough in himself to attract it, and also because he is utterly devoid of those finer sensibilities of the heart through whose promptings one is restrained from all acts of cruelty and torture, from all acts that will give pain to any living creature. I would point out to the child the torture that is inflicted upon the animal during the process of the sawing and the burning of the tail, and also that this acute pain and torture is but little compared with the after-torture that is to follow during the balance of the horse's life.

The skin of the horse is exceedingly sensitive to the bites and the stings of the flies and other pestiferous insects that harass him during the heated term of the year, and which without this natural weapon of defence make his life almost unendurable. I would point out to the child how cruelly the animal is maimed for life, and how foolhardily its beauty is forever destroyed.

The practice has already by statute been made a crime in a number of States, punishable by both fine and imprisonment, but still the idiotic, cruel, and deplorable practice goes on to a greater or less extent; and not until public sentiment is thoroughly aroused against it will it entirely cease. If the one who has it done were compelled to stand for but half a day in the hot summer weather, with his back bare to the bites and the stings of the flies and sweat-bees and other insects that would drive him almost frantic, if his hands were so fastened that he could not drive them away, then he might be brought partially at least to his senses.

And when the fine sensitive horse whose tail had been sawn off in this way, so that he was one day driven almost to madness by the stings and bites he was powerless to protect himself from, especially as he was farther maddened by that fiendish device of torture,

B

the high check-rein, finally became unmanageable and dashed down the road a runaway, hurling his owner to death and his wife to the bed of an invalid and cripple—it may seem unkind to say it—but it certainly served them right. They reaped only what they themselves had sown, as every one must in some form or another, for such is the law of the universe.

CATTLE TRANSPORT

And again, as an object-lesson, I would point out to the child the men who each year engage in cattle-starving on our Western plains; for on the various ranches thousands of head of cattle in cold winters starve and freeze to death, because left to themselves when they can no longer find sufficient food on the ranch, this plan being adopted by many cattle-raisers because it is cheaper for them to lose a certain portion of the herd each winter than it is to furnish them suitable food and shelter. Thousands of cattle have so perished during the past winter. I would show that such a man is a criminal and deserves restraint as such, no less than a man who would cause a part of his stock to starve to death in a stable or on a farm.

I would teach the child the same in regard to those responsible for the careless, cruel, mercenary methods of transporting cattle, sheep, and horses from the West to the East, or to England and other countries, in the cattle ships, where sometimes as many as a quarter or even a third of the animals are found dead on their arrival, and numbers of others so mangled and crippled that they have to be killed as soon as they are taken from the vessel.

DRESS AND FASHION

There is another excellent opportunity for humane teaching, and one that comes especially near to every woman. It lies in the thoughtless, cruel, and inexcusable practice of wearing the skins and plumage of birds for millinery and other decorative purposes. The enormous proportions of this traffic are simply appalling. In the course of a single day last year in London, and from a single auction store, the skins of six hundred thousand birds were sold. This number represented the sales of but one store of one city on a single day.

Millions of birds are destroyed annually to supply the demands that fashion venders, who become wealthy thereby, have created in the minds of women for this purpose. Whole species of birds have already become practically extinct by this wholesale slaughter, while others are rapidly becoming so. For example, that beautiful bird the white heron, commonly known as the egret,—in Florida but one can now be seen here and there by the tourist where thousands could be seen but a few years ago. This bird is killed and its plumage taken only at that season of the year when its dress becomes a little more brilliant than usual, for it is its nesting time, and Nature seems to be recognising this, the marriage season, by preparing for it its wedding garments.

The birds at this season are apparently very innocent of harm and very tame, and are found near together taking care of their young. At times hundreds of birds are to be found near together in one roost among the tall trees of the swamp-lands, so that the bird-catcher finds it an easy task to conceal himself and pick them off as they are return-

ing to their nests with food for their young,
—sometimes to the extent of several hundred
in a single day; and every bird killed at this
season means the starving to death, on the
average, of four or five of its young.   It be-
hoves every woman, then, who wears even a
single egret plume, to remember that she has
been the cause of the sacrifice of at least four
or five birds.

"But," says the gentle lady, "I had nothing to do with the killing of the birds."
True; had you to do with it personally you
would not wear what you now wear.   But
were it not for multitudes of ladies like yourself, Bill Jones, bird-catcher, would turn his
mind and energies to other avenues, for he
would no longer have a demand, and hence
a market, each year to supply.

I know of one bird-catcher who, with his
assistants, in a single season slaughtered and
took the skins of over one hundred and
thirty thousand birds.   Think what this
means when we take into consideration
the few days of the very short season devoted
to this!

And what does this indicate in women?
I would not be unfair, and so I will say that
to me it indicates chiefly thoughtlessness and

lack of imagination on her part. If the one who now decorates herself with the plumage of her slaughtered fellow-creatures could be on the spot with Bill Jones and see the crimson life-blood that the bleeding heart is pulsing out, staining even the feathers that she herself will wear—if she could see the agonies of the death struggle, and then see the gaping mouths of the starving young ones in the nest, waiting in vain for the return of the parent bird with food—then, I am sure, she would no longer be a victim to this foolish, thoughtless, heartless habit. No; I have too much respect for and faith in the finer sensibilities of woman to believe that she would. Once in a while, it is true, we will find a woman so wrapped up in her vain, selfish, insane desire for show that, notwithstanding the realisation on her part of all we have just said, she would nevertheless demand this sacrifice to minister to her vanity.

Were I a woman I certainly should want to be among the forerunners in the movement that has already begun along this line. I would rather be a leader in setting a good fashion than a follower of a poor and positively bad one.

And you will be surprised what beautiful

hats and bonnets can be devised by the woman of a little ingenuity, without the aid of birds' plumage or feathers of any kind. And when skilful minds and hands are once turned in this direction we shall wonder that this relic-of-barbarism mode of adornment, even though it be a somewhat modified form of it, has lasted so long.

As a mother I would keep or lead my daughter out of this heartless and needless practice by first abandoning it myself. Children are so quick to see inconsistencies. Said a little fellow to his mates the other day: "I know why teacher don't want us to rob the birds' nests and kill the little birds. She wants 'em to grow up so she can wear 'em on her bonnet." And when one sees, as I have seen, a teacher with the skins of two and the feathers of more birds on her hat, we will realise that, after all, teaching by example is better than by precept, or, putting it in another form, teaching by precept without its being reinforced by example is of but little value.

But for the people's sakes, as well as, if not even more than for the bird's, I would urge attention to and action along this line. The tender and humane passion in the human heart is too precious a quality to allow it to be hardened or effaced by practices such as we so often indulge in. Even from an economic standpoint, the service that birds render us every year, so far as vegetation is concerned, is literally beyond computation. Were they all killed off, the world would soon become practically uninhabitable for man, because vegetation each year would be so thoroughly blighted or even consumed by the hordes of insects that would infest it. It is but necessary to realise how rapidly, even during the past several years, insect life has been increasing in some quarters, so as to tax to the utmost the skill of the farmer, the gardener, and the fruit-grower. Instead, then, of schooling the child to be the destroyer of bird life, let it be guided along the lines of being its lover and its protector.

And if those who use them, women especially, could know and fully realise the cruel and at times almost unspeakable cruelty and torture that attends the procuring of their sealskin and other fur or fur-trimmed or lined garments, I am sure that many at least would begin quietly to look about for garments made of other materials;—if they could know of the seals being clubbed to death in their innocent tameness on their native ice rocks, of the other fur-bearing animals that are trapped, remaining for hours, or even at times for days, with leg or legs crushed between the trap's cruel and relentless steel jaws, before the merciful blow comes that is to end their torture, when it has not already died from its torture or from starvation, or has not gnawed its leg from the trap with its own teeth in order to escape—if they could be brought fully to realise these facts, then I am sure they would conclude that these articles are bought with a price greater than any *human* being can afford to pay.

FLESH AS FOOD

A word now in regard to another matter that is of far more importance than is generally supposed—the matter of the excessive flesh-eating that is continually going on in our country. After looking carefully into the matter, and after some years' experience in its non-use, I can state without hesitancy that, contrary to the prevailing opinion, the flesh of animals is not necessary as an article of food.   But few are better off for its use, while the great majority are the worse off for it, and especially is this true when it is so excessively used as we find it now on every hand.

We shall find numerous articles of food, as we study the matter, that, so far as body nourishing, building, and sustaining qualities are concerned, contain twice, and in some cases *over* twice, as much as any flesh food that can be mentioned.   The liability to mistake in this matter lies in the fact that flesh foods when taken into the stomach burn, oxygenise, more quickly than most other foods do, and this short stimulating effect, resembling more or less the stimulating effects of alcohol, is mistaken for a body nourishing and sustaining effect.

Flesh foods stimulate the passions, and more, acting as a stimulant in the body, they call for other stimulants to feed and satisfy the appetites thus aroused ; and some of the world's most eminent physicians, who have looked carefully into the matter, are declaring that the excessive amount of whisky and beer drinking, with its attendant drunkenness and crime, will never be

done away with, or materially lessened, so
long as this excessive eating of flesh con-
tinues.

Numerous other things, such as the irrit-
ability it causes in the natures of large
numbers of people who use it, the almost
unconscious blunting of many of the finer
senses, as also the dangers attending its use,
on account of the diseased or poisoned condition of meats in many cases, are worthy
of a very serious consideration. If space
permitted, many facts regarding the exceedingly large number of diseased animals that
are eventually sold in the form of meat,
facts as reported by various boards of in-
quiry, various commissions, etc., might be
cited; and who can tell when such may
not be the condition of that of which he
himself is partaking? And when we remember the vast numbers of animals—cattle
especially—that are angered almost to de-
speration, in some cases literally maddened
by anger, and when we remember the
peculiar poison that reaches every part of
the body when the mind is thus angered,
and that in this state large numbers of
animals are killed, we can readily see how
important this aspect of the matter is.

"But is not flesh-eating natural?" I hear it asked. "Does not man in his primitive, savage state make use of flesh *naturally*? Do not animals devour one another?" Yes; but we are not savages, nor are we purely animals, and it is time for us to have outgrown this attendant-of-savage-life custom. The truth of the matter is that considerably more than the one-half of the people in the world to-day are not flesh-eaters. And many peoples, whom large numbers in America and in England, for example, refer to as the heathen, and send missionaries to Christianise, are far ahead of us, and hence *more Christian* in this matter. And one reason why missionaries in many parts of India, among the Buddhists and Brahmins, for example, have been so comparatively unsuccessful in their work is because the majority of those keen-minded and spiritually unfolded people cannot see what superiority there is in the religion of the one whom it allows to kill, cook, and feast upon the bodies of his or her fellow-creatures, which they themselves could not do.

In Bombay, to have the carcasses of animals exposed to public view, as we see them in the stores and markets here, and at times scores of them decorating their windows and entire fronts, is prohibited by law.

No, experience will teach you that if you do away with flesh-eating and get in its place the other *valuable* foods, the time will quickly come when you will care less and less for it; then again, the time will come when you will have no desire for it, and finally, you will grow positively to dislike it and its effects, and nothing could induce you to return again to the flesh-pots. And as for those who think that the ones who are not flesh - eaters are necessarily weaklings, I should like to match a friend of mine, an instructor in one of our great American universities, who for over eighteen years has eaten no flesh foods,—I should like to match him with any whom they may send forward, when it comes to a test of long-continued work and endurance.

In London there are already numbers of restaurants where no flesh foods are served; in Berlin there are already about twenty, and their number in these, as well as in numerous other cities, is continually increasing. It is a matter of but a short time when there will be numbers of such in our own country. The only really consistent humanitarian is the one who is not a flesh-eater; and great, I am satisfied, will be the results, both to the human family and to the animal race, as children are wisely taught and judiciously directed along this line.

When one goes into the better restaurants where no flesh foods are served, in England and Germany for example, he is impressed with the foundationless excuse of so many people, that it is hard, or even impossible, to get along without flesh foods. In the other realms will be found an abundance, a hundred or a thousand times over, and especially when we begin to give some little attention to the great varieties of most valuable foods there, and to the exceedingly appetising ways in which they can be prepared. One reason why such large numbers of people feel that meat is a necessity, or almost a necessity with them as an article of food, is because in our hotels and restaurants and cafés, and, in fact, in the majority of our homes, the meat element forms the chief portion of the foods prepared for our tables, and to it,

practically, all the skill in preparation is
given; while the other things are looked
upon more as accessories, and are many
times prepared in an exceedingly careless
manner, much as mere accessories would be.
But with a decreasing use of flesh foods
and with more attention given to the
skilful preparation of the large numbers of
other still more valuable foods, we shall
begin to wonder why we have so long been
slaves to a mere custom, thinking it a
necessity.

An eminent Hindu has presented some
truths along the lines of non-flesh eating
so ably, that I yield to the impulse to
quote from him quite at length:

"Animal flesh enriches the blood with unnecessary *fibrin*, and this produces unnatural
heat in the system, and in turn is the cause of
unusual activity and restlessness, ultimately leading to the nervous debility which afflicts many
meat eaters. Constant use of meat increases
the action of the heart and brings premature
loss of vital forces. Physiologists and compara-
tive anatomists like Sir Everard Home have
shown from the structure of the teeth, stomach,
alimentary canal, the microscopic human blood-
corpuscles and the digestive processes, that man

is by nature more related to frugivorous animals
than to the carnivora.

"From the chemical analysis of different
vegetables, cereals, fruits, nuts, etc., and the
flesh of different animals, and from the com-
parison of the constituent properties of vegetables with those of animal flesh, it can be shown
that everything necessary for the growth of the
muscles, for the strength of the nerves, and for
the nourishment of the whole body can easily
be obtained from the vegetable kingdom. This
being the fact, the question arises, Why do we
eat animal flesh? Is it for nourishment? No.
The same nourishment can be obtained from
vegetables, cereals, and pulses. Is it for health
that we eat meat? No; because vegetarians,
as a class, are healthier than the majority of
meat-eaters. Why, then, is meat eaten? Because of the habit transmitted from generation
to generation, and because of superstition, pre-
judice and ignorance.

"Various objections have been raised by meat
eaters against vegetarianism. Some say if animals are not used for food they will overrun the
earth. In India the Hindus do not kill cows,
but they are not overrun by them. The Hindus
did not have any slaughter-houses until the
British Government established them. In the
States that are still governed by the Hindu
Râjâs the wild animals and birds are protected
by strict laws. But these States are not over-

run by wild animals, nor are the inhabitants driven out by them. Others hold that unless they eat animal flesh they will be weak and useless for work and will lack bravery and courage. This is a great mistake. You have heard of the Hindu Sikh soldiers in India, who are the bravest and strongest fighters in the British army. They never turn their back to an enemy in the battlefield. One Sikh soldier can stand against three beef-eaters in hand-to-hand fight. But these soldiers never touch meat, nor fish, never drink wine, nor smoke tobacco. They are strict vegetarians. A vegetarian diet gives great endurance and makes one even-tempered. People generally mistake a ferocious, restless, and rash temper for courage and strength. These say that a tiger or a wolf is stronger than a horse, a buffalo or an elephant. They make ferocious nature the standard of strength. It is true that a tiger can kill a horse, but has he the muscular strength which enables a horse to draw a heavy load a long distance? A tiger can kill an elephant, but can he lift a cannon weighing hundreds of pounds? Ferocity is one thing and muscular strength is another: we ought to distinguish the one from the other. The source of strength lies in the vegetable kingdom and not in flesh and blood. If flesh eating be the condition of physical strength, why do meat-eaters prefer the flesh of herbivorous animals and not that of the car-

c

nivora? Some meat-eaters say that animal flesh has a large quantity of vegetable energy concentrated in a small compass. If that be their reason for the meat - eating habit, they ought to live on the flesh of carnivorous animals and birds, such as tigers, wolves, vultures, and hawks.

"As in the animal kingdom the carnivora are more restless than the herbivora, so amongst men we find that meat-eaters are more restless and less self-controlled than vegetarians. As a peaceful, poised and self-controlled nature is the first sign of spiritual progress, it is plain that animal food is not the most helpful diet for spiritual development."

The time will come in the world's history, and a movement is setting in that direction even now, when it will be deemed as strange a thing to find a man or a woman who eats flesh as food, as it is now to find a man or a woman who refrains from eating it. And personally, I share the belief with many others, that the *highest* mental, physical, and spiritual excellence will come to a person only when, among other things, he refrains from a flesh and blood diet.

Personally, I shall be glad, as long as forces and agencies are at work that tend to keep armies in the field, if we awaken, and that almost instantly, to the dangers attending the health of troops in service from the large amount of " canned meats " that are used in connection with army rations. I believe, and I think I should be fully borne out by the facts if they could be thoroughly known, that thousands of deaths due to disease have been to a great extent induced or helped on through this agency. The evidence brought out in the investigations along this line, in connection with the American forces which served but recently in the Spanish-American war —the conclusions of which were presented to the people as skilfully as possible, and were then allowed to drop as quickly as possible—should have no small weight with us as a people. If it were to receive the attention it really demands, many thousands of lives might be saved in the future that otherwise may be needlessly sacrificed.

Were such a food necessary, it would then be a different matter ; but when there are other foods, even more valuable so far as body-building and nourishing and sustaining qualities are concerned, and more free from the poisoned and loathsome conditions that so much of the canned meats get into, especially in hot climates—foods that can be transported just as readily; in fact, prepared in a similar way and ready for immediate use—then we can readily see the criminal folly in allowing a continuance of its use, at least in such quantities as it is at present used.

And there is another matter of grave importance that we should not be allowed to lose sight of in this connection. The brutality to the animal creation, which as a weaker creation we should protect and care for, has its corresponding and balancing element in connection with our duty to those who are hired to do our butchery for us. And here let me quote a paragraph from Mr Henry Salt, the well-known English humanitarian thinker and worker:

" But this question of butchery is not merely one of kindness or unkindness to animals, for by the very facts of the case it is a *human* question of no slight importance, affecting as it does the social and moral welfare of those more immediately concerned in it. Of all recognised occupations by which, in civilised countries, a livelihood is sought and obtained, the work which is looked upon with the greatest loathing (next to the hangman's) is that of the butcher, as witness the opprobrious sense which the word 'butcher' has acquired. Owing to the instinctive horror of bloodshed which is characteristic of all normal civilised beings, the trade of doing to death countless numbers of inoffensive and highly organised creatures, amid scenes of indescribable filth and ferocity, is delegated to a pariah class of 'slaughter-men,' who are thus themselves made the victims of a grievous social wrong. 'I'm only doing your dirty work; it's such as *you* makes such as *us*,' is said to have been the remark of a Whitechapel butcher to a flesh-eating gentle-man who remonstrated with him for his brutality; and the remark was a perfectly just one. To demand a product which can only be procured at the cost of the intense suffering of the animal, and the deep degradation of the butcher, and by a process which not one flesh-eater in a hundred would himself, under any circumstances, perform, or even witness, is conduct as callous, selfish, and unsocial as could well be imagined. . . . To have accustomed one's self to a total disregard for the pleading terror of sensitive animals, and to a murderous use of the knife is a terrible power for society to put into the hands of its lowest and least responsible members. The blame must ultimately fall on society itself, and not on the individual slaughterman."

Chicago has gained, temporarily, at least,

the reputation of being the great slaughter-city of the world. Some of Chicago's first officials in the police department have given us many facts showing the direct connection between the influence of this trade, or rather this "business," and some of the most shocking crimes that the city has known of late years, for large numbers of these have been perpetrated by men engaged in this business, who have been gradually reaping the deteriorating effects of its influence.

"No one who goes to Chicago," says a writer in the *New Age*, "should fail to see the shambles. They are the most wicked things in creation ; they are sickening beyond description. The men in them are more brutes than the animals they slaughter. Missions and institutes have been built in respectable parts of the cities from the profits, and the employees themselves have been left to go straight down to the devil. . . . It is the duty of everyone interested in social questions, of everyone whose demands necessitate this kind of labour, to wade through this filth to see those poor wretches at work."

One who visited one of the Kansas city packing-houses, where many thousands of animals are killed daily, and where some thousands of men and boys are employed, writes as follows :—

" Inside the vast slaughter-house it looked like a battle-field—the floors were crimson ; the men were deep-dyed from head to foot. It was a sickening spectacle. There the cattle were driven into pens, scores at a time, and the echo of the pole-axes was heard like the riveting of plates in a ship-building yard. Then the gate fronts were raised, and the kicking animals were shot on to the floor, to be seized by the hoofs by chains, and hoisted to the ceiling, and sent flying on their way to rows of men, who waited with knives, and skinned and quartered and washed them."

In the light of the foregoing facts, and in the light of many more that might be presented, we can readily see that each one who aids in creating the demand for flesh foods is to a greater or less extent, not indirectly but directly, responsible for the degrading and dehumanising influences at work in the lives of many thousands of their fellow-men. We *are* our brother's keeper whenever it comes to a matter that we are personally involved in, and there are responsibilities that we cannot shift after we are once made acquainted with the facts pertaining to them.

May I present here a few additional thoughts along this line, given utterance to by that very clear-thinking woman, Annie Besant:

"We may adopt a bloodless diet to purify the body, or in order that we may have a body that will be less an obstacle to intellectual and moral growth ; and such reasons as these justify the practice, and no man or woman need be ashamed to confess them. But still deeper and more attractive than such an object is our *principle*, our recognition of the *unity* of life in all that is around us, and that we are but parts of that one universal life. When we recognise that unity of all living things, then at once arises the question—How can we support this life of ours with least injury to the lives around us ? How can we prevent our own life adding to the suffering of the world in which we live? . . . And at once we begin to see that, in our relations to the animal kingdom, a duty arises which all thoughtful and compassionate minds should recognise — the duty that, because we are stronger in mind than the animals, we are, or ought to be, their *guardians* and *helpers*, not their *tyrants* and *oppressors*, and we have no right to cause them suffering and terror merely for the gratification of the palate, merely for an added luxury to our own lives.

" . . . Thus looking upon the animal kingdom, a sense of duty awakens within us ; we feel that they are not intended simply to be slaves of men's whims, to be victims of his fancies and desires ; they are *living creatures*, showing forth a Divine life, in lesser measure than ourselves, it may be, but it is the same Divine life that is the heart of their heart, and the soul of their soul.

"The animal evolves under the fostering intelligence of man. The horse, the bullock, the dog, the elephant, any of the creatures that are around us in different lands, all develop a growing intelligence as they come into healthful relations with their elder brethren, men and women. We find that they answer with love to our love, and also with growing intelligence ; and we begin to realise that it is our duty to train and help that growth by making them co-workers with ourselves, to develop their intelligence by human companionship ; and not to slaughter them and thus make a gulf of blood between them and mankind.

"Surely man should not go through nature leaving behind him a track of destruction, of misery, of hideous injury.

" . . . So that one standpoint we may take up as Food Reformers is the standpoint of

Love, of recognition of our true place in the world. Not only that we may have cleaner materials in our bodies, not only that we may have a better instrument for our minds and souls to work with, but that *we may be better channels of Divine Love to the world on every side*. For this reason, fundamentally, I am a vegetarian, and I would not take for myself, needlessly, the life of any sentient creature that lives around me.

" . . . But no one can eat the flesh of a slaughtered animal without having used the hand of a man as slaughterer. Suppose that we had to kill for ourselves the creatures whose bodies we would fain have upon our table, is there one woman in a hundred who would go to the slaughter-house to slay the bullock, the calf, the sheep, or the pig? Nay, is there one in a hundred who would not shrink from going to see it done, who would not be horrified to stand ankle deep in blood, and see the carcasses lying there just after the animals were slain? But if we could not do it, nor see it done; if we are so *refined* that we cannot allow close contact between ourselves and the butchers who furnish this food; if we feel that they are so coarsened by their trade that their very bodies are made repulsive by the constant contact of the blood with which they must be continually be-smirched; if we recognise the physical coarseness which results inevitably from such contact, dare we call ourselves refined *if we purchase our refinement* by the brutalisation of *others*, and demand that *some* should be *brutal* in order that *we* may eat the results of their brutality? We are not free from the brutalising results of that trade *simply because we take no direct part in it*.

" . . . And everyone who eats flesh meat has part in that brutalisation ; everyone who uses what they provide *is guilty of this degradation of his fellow-men*.

" . . . I ask you to recognise your duty as men and women, who should *raise* the Race, not *degrade* it ; who should try to make it *divine*, not *brutal* ; who should try to make it *pure*, not *foul* ; and therefore, in the name of Human Brotherhood, I appeal to you to leave your own tables free from the stain of blood, and your consciences free from the degradation of your fellow-men."

If the one who uses *pâté de foie gras*, *ortolan*, and other abnormally formed things of this type, will look a little into the methods by which they are obtained, with all their agonising and slowly dying torture attendants, I dare say he will then use them no longer ; he will not, indeed, if there is in him a nature that can truly be described by the word human in distinc-

tion from that of brutal. It is our thoughts
and our acts, or our complicity in the acts
of others, that determine whether at any
given time we are nearest akin to the brute
or the human.

It happened not long ago, in looking over
the advertising pages of one of our great
monthly magazines, that my attention was
called to a whole-page advertisement of one
of Chicago's packing-houses. In connection
with this advertisement numbers of figures
were given, among which were the following :—

"Six packing-houses, sixty-five acres of
buildings. Handled last year, 1,437,844
cattle, 2,658,951 sheep, 3,928,659 hogs.
18,433 employees."

Here, then, is a total of a little over
8,000,000 animals slaughtered in a single
year by one concern, and when we take into
consideration the number of other concerns
of similar magnitude, and also the thousands
of other slaughter-houses in the various cities
and villages throughout the country, we can
perhaps form at least some idea of the vast
proportions of this traffic in blood. And
when we take into consideration that in this
one concern over 18,000 people were em-

ployed, we can also form some slight con-
ception of the large number of men, women,
and children throughout the country who are
brought under the influences which we have
been considering.

And then, by a strange coincidence, though
the connection is natural, I turned a page or
two and my eye fell upon the advertisement,
also a full-page advertisement, of a large
brewing concern.   The advertisement in part
ran as follows :—

"When 219 carloads of —— Beer were
shipped to Manila the world wondered. What
industry was this that shipped its product
by a mile and a half of trains to that re-
mote spot?

"Yet that enterprise has been repeated a
hundred times over.   Wherever civilisation
has gone —— Beer has followed.   Agencies
have for twenty years been established in
many of the farthest parts of the earth.

"—— Beer has been known in South
Africa since the white man went there.   It is
shipped in large quantities to the frigid wilds
of Siberia.   It is advertised in the quaint
newspapers of China and Japan.   It is the
beer of India, the beverage of the Egyptian
and the Turk.

" It is too little to say that the sun never sets on —— agencies, for it is literally true that it is always noonday at one of them."

Marvellous indeed is the enterprise of the great nation, so magnificently equipped for carrying, among other things, a flesh and blood, a whisky and beer civilisation, which we complacently denominate by the term Christian, to the remotest parts of the earth and to the benighted peoples who stand so in need of these "civilising" influences; peoples, moreover, who are so obtuse and so "stubborn" in the face of their benevolent Anglo-Saxon well-wishers, that many times these civilising influences can enter only after the blood of numbers of their bravest, most highly educated and patriotic sons has been spilled, through the agency of rifles, bayonets, and Gatling guns.

SPORT AND WAR

It does not require any great amount of mental power to be able to trace the direct connection between a spirit of kindness and consideration and care for the animal world and a kindred spirit of kindness, care, and

consideration for all of the human kind,
as also between this and a tendency to settle
differences of opinion, or disputes, by the
thoroughly wise and economical method of
conciliation and arbitration in distinction
from the thoroughly unwise, expensive, and
degrading method of swagger and bravado,
which leads so often to a resort to force
in the form of individual, or corporate, or
national murder.

There is a direct connection between
"pig-sticking" and man-bayoneting. There
is a direct connection between the foremost
representatives of a great nation, and a large
class that ape and follow them, in shooting
hundreds of pheasants or partridges in a
single day, and a spirit of reckless bravado
that easily leads the nation into war with
another nation when complications arise, or
when, through the agency of superior force,
it can gain its ends in appropriating to itself
the wealth of the goldfields or the territory
of a smaller or weaker people.

There is a direct connection between an
Emperor's proclivity for killing that leads
him and his party ruthlessly and flippantly to
shoot two or three thousand pheasants in a
single day and to pride themselves upon the

achievement, and words such as were recently addressed to a portion of the army at the head of which he temporarily is, enjoining them, above all things, to go out with *revenge* as their watchword, on account of certain indignities shown to a few citizens of that country in the far East recently. The spectacle of one who prides himself upon being the head of a *Christian* nation, counselling and arousing and fostering a spirit of revenge is indeed anomalous. But the step is not a long one from blood-spilling in the animal world to the same in connection with human beings, and to a spirit of hatred and revenge which shows that most essential elements of a *true Christian* character are wanting.

There is a direct connection between the unwise, expensive, and thoroughly unstatesmanlike method of dealing with the recent differences in South Africa, and the Royal Buckhounds which to-day ornament—no, not ornament, but disgrace—the Queen's Court. The time is coming when the practice of tame deer hunting will be as much looked down upon, and condemned as brutal and unworthy English gentlemen, as the bull- and bear- baiting that prevailed so universally in Elizabeth's day are at the present time.

The time is coming when in England, in Germany, in America, in every nation, the people will find that there is a higher duty than that of following the leadership of men miscalled statesmen, because lacking in that spirit of honest, frank consideration and conciliation, through lack of which disputes are allowed to come to a settlement by force of arms, the consequent burden being thrown upon the people to bear. The time is coming when we shall find that this is not patriotism, but that patriotism is that ready service which works for the people's highest welfare, and that the people's highest welfare is served most by keeping their country, among other things, out of expensive and demoralising bloody warfare, rather than by getting it into war. It was the President of the American Humane Education Society who most truthfully said : " *Unnecessary* wars are simply wholesale murder, and the men who cause them [however high their positions] are the greatest and worst of criminals."

It is the same spirit of killing and "pig-sticking," and looking upon the animal

D

world simply as something to use for our own pleasure or gain, without any consideration of their rights, and without any impulse to care for and treat them kindly, that fosters that thoroughly insane imperialistic tendency that has gained such footing in England of late, endangering the very existence of the empire, and demanding the enormous price to be paid that is being demanded to-day. The same imperialistic tendency in America has lately brought the nation dangerously near to the parting of the ways, one of which leads to the continual upbuilding of a republic, the pride of all time, the other to its gradual undermining, by transforming it into an empire — if not in name, in reality — and thus sending it, through the violation of great *elemental laws*, to the same ends that all nations that have adopted a similar course have or must inevitably come to.* It is this spirit, a tendency to which has been witnessed in America but recently, which will gradually blunt, and in time entirely kill, the quick, noble, and God-like expression of sympathy for a people struggling for freedom and liberty. The destiny

* There is a world wide difference between territorial and national expansion in accordance with just and righteous laws and an aggressive and in time self-destructive imperialism.

and power of the American nation depends upon the fostering of the spirit of kindliness, love of fair play, and desire for conciliation, and hence of peace, in distinction from the spirit of militarism which great corporate interests and corporate politicians would have dominant in the country.

When the interests of the people are zealously guarded and righteously cared for, then when an emergency arises and there is a call to arms for defensive purposes, the only possible justification of any resort to arms, the nation will find that it has a citizen soldiery as vast as the numbers of its male population, and far more effective in the long run, than any hired body of soldiery can possibly ever be. And instead of supporting a vast army of men who are producing nothing, contributing nothing to the nation's welfare, but living upon the labour of others, and waiting merely for orders to shoot, mangle, and do to death men, fellow-men of another nation, each will be working for himself and for all, and will be enjoying the fruits of his labour. It is to humane education that we must look to save us from the monstrous system of militarism which at present prevails in the larger share of European countries.

In passing through Germany not long since, I was particularly impressed with seeing in fields here and there large companies of soldiers drilling and manœuvring, while in the fields on all sides of them, were numbers chiefly of women and children, and oxen and horses hard at work. This condition prevails to a great extent over the greater part of Germany. It is so, to a greater or less degree, in the other European nations where the military system has grown to such enormous proportions.

To seek neither the gold fields nor the territory of other peoples, to live in peace with all nations so far as in us lies, to be willing in all cases to give justice, as we are quick to demand it, to believe thoroughly that there are no questions or complications arising that cannot be settled by conciliation and arbitration if there is the earnest sincere desire to do so, and to have in public office men who are imbued with this idea, refusing admittance to those who are not wise enough to be guided by this principle, but in whom the spirit of swagger, brow-beating, and bravado prevails—in these, among other things, lies the hope, the healthfulness, the great and growing power, the future grandeur of the American nation. It is in this way that she can preserve and maintain and continually increase her unique position among the nations of the earth.

This reign of peace is indeed the condition that all people who are humanely inclined, all people who are lovers of animals, should work to bring about, and thus to save the many thousands of horses and mules and oxen the inhuman treatment and the terrific suffering to which they are always subjected when a war is in progress. The treatment that thousands of animals have been subjected to both on transport and on train, and on the field in South Africa during the past few months, is a burning disgrace to the British nation. Brutalities have been engaged in and condoned, that would not be countenanced for an instant by the government of this or any other nation at all civilised, in ordinary circumstances. The very nature of the conditions, of course, makes it hard for the noble and willing animals to be carefully attended to and mercifully treated. And this is greatly accentuated by the fact that every diabolical agency is let loose which increases the spirit that actuates the ill-treatment and the most awful abuses on the part of those who have the animals in charge. But this very fact makes it all the more imperative for those who value humane education, to work all the more zealously and unceasingly for its universal advancement, so that conditions of this kind may be prevented, and the causes of a terrible amount of suffering to hosts of noble animals may be done away with.

The time *is* coming when practically all, with Cowper, will say and feel:

"I would not enter on my list of friends,
  Though graced with polished manners and
      fine sense,
  Yet wanting in sensibility, the man
  Who needlessly sets foot upon a worm."

This is our ultimate destiny, though we have been coming up the steep most tardily.

Personally, I would rather be the author, and have the rare unfoldment of heart and sympathy of the author, of the following little stanzas, than be the greatest military leader in the world to-day :—

"Across the narrow beach we flit,
  One little sandpiper and I ;
And fast I gather, bit by bit,
  The scattered driftwood bleached and dry.
The wild waves reach their hands for it,
  The wild wind raves, the tide runs high,
As up and down the beach we flit,—
  One little sandpiper and I.

I watch him as he skims along,
  Uttering his faint and mournful cry ;
He starts not at my fitful song,
  Or flash of fluttering drapery ;
He has no thought of any wrong,
  He scans me with a fearless eye,—
Staunch friends are we, well-tried and strong,
  The little sandpiper and I.

Comrade, where wilt thou be to-night,
  When the loosed storm breaks furiously?
My driftwood fire will burn so bright !
  To what warm shelter canst thou fly?
I do not fear for thee though wroth
  The tempest rushes through the sky ;
For are we not God's children both,
  Thou, little sandpiper, and I."

Instead of the spirit of destruction and
the desire to gain something for ourselves,
to kill something, to tear something from
its life, even at the expense of breaking up

that wonderful harmony which reigns in God's world, we need the spirit which animated Emerson when he wrote the lines entitled "Forbearance":

"Hast thou named all the birds without a gun?
  Loved the wood-rose, and left it on its stalk?
  At rich men's tables eaten bread and pulse?
  Unarmed, faced danger with a heart of trust?
  And loved so well a high behaviour,
  In man or maid, that thou from speech re-
    frained,
  Nobility more nobly to repay?
  O, be my friend, and teach me to be thine!"

TREATMENT OF CRIMINALS

In the degree that moral, heart, humane training finds its place among us as a people, in that degree shall we come nearer a wise and humane method of treatment, so far as the more unfortunate ones among us, whom we denominate by the term "criminal," is concerned. It is a well-known fact that we have not as yet found the true method of dealing with these our fellow-beings. Our methods deal too much with punishment, and not enough with unfoldment, and there-by prevention. Our methods in the long run tend to make criminals, and to per-

petuate criminals, rather than to prevent
them or to transform them into law-abiding
and honourable citizens. When a man
makes a mistake that any of us might have
made, and that possibly under like conditions many of us would have made, the
spirit of punishment for the sake of punishment, even to the extent of revenge, so
holds us as a people that we truly share
in the wrong-doing of the one whom we
condemn and injure, when by another,
a more sane, a more thoughtful, a more
kindly and common-sense method, we
would be instrumental in bringing about
a set of conditions which, instead of perpetuating the man as a criminal, would
make him an honour and a blessing to the
community in which he lives.

Likewise, when through misfortune or
broken health, an inability to find work,
and many times in a starving condition,
a man or a woman is compelled to find
entrance to the workhouses in England,
in many at least he is treated more as a
beast, or as an inanimate object, than as
a human being. He who enters these must,
as a rule, leave all hopes for love and kindly
and sympathetic treatment behind. And

this, indeed, is the reason why so many deliberately take their own lives rather than enter them. And yet it is England who prides herself upon being the world's greatest Empire—a great Christian nation whose mission it is, even with shot and shell her leaders will tell you, to carry the blessings of a Christian civilisation to the inferior peoples of the world.

When we once begin to understand that ignorance is at the bottom of all wrong-doing, of all sin and error and crime, with their attendant sufferings and losses, then we shall begin to realise that sympathy and compassion—and, consequently, kindly treatment — instead of punishment and revenge, is necessary if we would truly aid one who has stumbled. The systems in vogue to-day will make and will perpetuate criminals; they will not transform a wrong-doer into a strong, sympathetic, and honest man or woman. They may make him or her a greater danger to society, but they will never make him or her an aid to the community, or an aid in bringing about a higher state of life and civilisation in the community, as practically every one of such can be made. In the work of the

George Junior Republic, in New York State, we have an evidence of what can be accomplished when work is begun from the right side. And if boys can be reached so effectively by these methods, certainly men and women can be also.

During a single year recently a hundred and sixty thousand cases in round numbers were committed to prison in England and Wales. Of this number over sixty-one thousand—considerably more than a third—were committed for a week or less. Now, under a wise and more enlightened system of penal law, most of these cases need not, and should not, have come to prison at all. In the matter of imprisonment it is so often that it is the first step, the first commitment, which counts and which eventually evolves a criminal future. Of these sixty-one thousand cases and over, a very large number were first offenders, and many were imprisoned for but three or four days. How often it has been said by one whose life has been one of the criminal cast, "If I hadn't got that three or four days (or that week) when a boy, how different my whole life would have been." If, therefore, we would prevent having a criminal class we must do everything in our power to prevent unnecessary additions to it, and especially should we refrain from actually driving early and slight offenders into its ranks.

And then the meaningless, unnecessary, and deplorable degradations that so often accompany the treatment of prisoners are worthy of the most unreserved condemnation. There is enough degradation, God knows, accompanying the entrance to prison life in itself without any studied additions to it. When, through studied efforts, or the blind and brainless following of bad precedent on the part of prison officials, it is made next to impossible for one in prison to receive frank and open-hearted kindness, and when thereby it is made impossible for him spontaneously to give kindness, then no more successful steps in the process of perpetuating him as a criminal, could be taken. Instead of wise, Christ-like steps to awaken, to feed, and to foster this greatest of heart qualities, hand-in-hand with self-respect, from the very first, crushing blows are given for its total destruction. Little wonder is it, then, that so often the offender comes out of prison with that deep and sullen hatred of all established order, that makes him more dangerous to himself and to society at large than ever before.

Any system of penal law and prison discipline or treatment that does not give a man or a woman back to the world better than when he or she entered the prison gates, is one greatly to be deplored. Here and there, however, there are brave and able men and women, strong, sweet, and with great love in their natures, who are giving themselves to this work, and who are quietly and gradually leading us into a better day. And as better social and more equal industrial conditions for the great mass of the people come about, and as a more vital, humane, heart-training for all classes, from the so-called highest to the so-called lowest, takes its place amongst us, then this great and constant problem will be already to a great extent solved.

We need more sympathy in all of our relations in every-day life, individual and national, and any methods of punishment that have in them the elements of resentment and revenge, in distinction from being restraining, educational, and uplifting are thoroughly anti-Christian, to say nothing of their being unwise, inexpedient, and expensive. Who shall accuse and who shall condemn? Certainly no wise man or woman; and certainly the unfortunate ones among us should not be in the hands of those who are the unwise. The time was, not so very long ago, when the insane were treated much as our criminals are treated to-day, treated as if they were to blame. A wiser spirit, however, prevails in regard to this unfortunate class among us, and insanity is now looked upon as a form of mental disease, not as a wilful perversion of one's natural self.

The wiser among us, who have given time and attention to the study of the criminal classes and the best methods of aiding them, are recognising that there is such a thing as moral disease, just as we have come into the realisation of the fact that there is such a thing as mental disease, and when those whom we call criminals are treated in accordance with these facts, then we shall begin to witness a great change for the better in our present methods.

Sympathy must be brought about so far as our relations with one another and so far as our relations with the animal world are concerned. Every living creature must be looked upon, respected, and treated as a living creature and not as a mere thing, not as something that is merely to serve our own purposes, with no right of any claims upon us in return.

Do you know the story of "The Caged Thrush"? A stanza comes to my mind:

"Alas for the bird who was born to sing!
They have made him a cage; they have
    clipped his wing;
They have shut him up in a dingy street,
And they praise his singing and call it sweet;
But his heart and his song are saddened and
    filled
With the woods and the nest he never will
    build,
And the wild young dawn coming into the
    tree,
And the mate that never his mate will be;
And day by day, when his notes are heard,
They freshen the street, but—alas for the
    bird!"

The Golden Rule must be applied in our relations with the animal world just as it must be applied in our relations with our fellow-men, and no one can be a Christian man or woman, or even truly deserve the

name of man or woman, until this finds em-
bodiment in his or her life.   Even worms are
our helpers, and it would be absolutely impossible, so far as the right conditions in the
ground are concerned, to get along without
them.   We are their debtors to a vast extent,
and were it not for the birds, practically
all vegetable and plant life would in time,
as we have found, be destroyed, and we
would be helpless even so far as our very
existence is concerned.   When we study the
habits of animals in a truly sympathetic
way and become thoroughly acquainted with
them and with the work that each one is
performing, we shall see that each one has
its place in the economy of God's world, that
each has its part to play, and that even so
far as the animal world is concerned we are
all related and inter-related.   If we destroy
or permit to be destroyed that marvellous
balance which the Divine Power has insti-
tuted in the Universe, we do it at our own
peril.   Instead, then, of being the enemies of
the animal world, instead of being its perse-
cutors and its destroyers, we should be its
friends and helpers.

HOMES FOR ANIMALS

Much among us is done for man, little as yet for the animal.   There are among us almost innumerable hospitals and homes for men and women, but there is very little of this nature as yet for the animals.   As yet, there is a Home or a Rescue League for animals only here and there.   We need them more abundantly.   We need Homes and Rescue Leagues and Clinics and Hospitals for them as we need them for ourselves ; and where there is one Animal Home to-day, there will be, I am sure, scores, or even hundreds, in time to come.

In far-off Bombay is probably the largest and most elaborate hospital for animals in the world.   It has both its in-patients and its out-patients, and it ministers to animals of all kinds as carefully as human beings are administered to in the hospitals of the West.   Over 2000 animals are taken into the hospital each year, and well on to 1000 are treated as out-patients.   In all there are some forty buildings, large and small, connected with the hospital, and the architectural structure and the appointments of some

E

of them are indeed superior to those of many of our regular hospitals.

This splendid hospital for animals was founded by a native Indian, a Parsee merchant, Sir Dinshaw Manockjee Petit. It is called Bai Sakarbai Dinshaw Petit Hospital for Animals, and receives its support from large numbers of citizens of Bombay who are interested in its beneficent work.

Not only domestic animals of every kind are treated and cared for in it, but the animals of the jungle and the wild birds which are found wounded or suffering from any cause, are taken to it and nursed back to health and then set free again.

The hospital is the pride of Bombay, and the Hindus are very liberal in their contributions to it. When endowing the laboratory Sir Dinshaw made the express stipulation that no vivisection should be practised in it, "for the reason that the same would wound the feelings of Hindus, from whom material support is obtained for the hospital, and if they come to know of it they will at once discontinue their support, and the hospital will thereby suffer in this respect." This is the frank and child-like reason given by Sir Dinshaw in one of the sections of the document by which he created the trust for the laboratory.

In addition to this splendid Hospital for Animals, there is in Bombay an influential Society for the Prevention of Cruelty to Animals. There is also the Pinjrapole, a place where worn-out or diseased animals are sent to be cared for until they are restored to health or until they die. Near Calcutta there is also a similar institution, established some thirteen years ago by a society of influential Hindus. It is near the Sodepur Station, some ten miles from the city, and is under the control of a manager with a staff of some eighty helpers and experienced veterinary surgeons.

In many cities in India institutions similar to those above described are to be found. Says a writer in the *London Telegraph* in describing this home for animals in Calcutta :

"It is true that the mysterious lower world of animal life is regarded in India with more reverence and kindliness than among Christian peoples. The one great fact of abstinence from flesh food produces an extraordinary effect among Hindoo communities. A newly-arrived

European walking in Baroda, or Nassick, or
any such Brahmanic capital, would mark with
wonder how the lower creatures have under-
stood and acted upon this tacit compact of
peace.   In the densest portions of the towns
the monkeys sit and chatter on the roof ridges,
the striped squirrels race up and down the
shop poles, the green parrots fly screaming
about the streets, the doves perch and coo and
nest everywhere, the flying foxes hang over the
most frequented wells and tanks, the mon-
goose scurries in and out of the garden gates,
the kites and crows frequent the market-places,
jungle doves and birds of all sorts forage boldly
for food, and at night even the jackals steal
impudently down into the suburbs.   There is
a great fixed peace between man and his in-
feriors in the scale of creation, and the effect
of this, to any lover of nature, is certainly
charming."

Here let me quote a few sentences from
a Hindu writer and teacher, personally
known to and honoured by many in America
and in England:

"When Hindu boys and girls go to school
and read their first lessons, they learn the
highest humanitarian principles, and as they
grow older they are kind toward all living
creatures.   They are taught: 'Be kind to
lower animals.   Do not kill them for your food,

because the natural food of man is not an animal.' I learned in the first book of Sanskrit: 'When enough of nourishment can easily be obtained from that which grows spontaneously on the earth, who will commit such a great sin as to kill animals for filling his stomach and deriving a little pleasure of taste?'

"Each one of these animals possesses a soul, has individuality and the sense of 'I,' can feel pleasure and pain, has fear of death and struggles to live. The germ of life in each one of these will gradually pass through the various stages of evolution, and ultimately appear in a human form. Therefore, the religion, philosophy, and Scriptures of the Hindus teach that as life is dear to us, so is it dear to the lower animals; as we do not wish to be killed, so they too shrink from death. 'Do not kill any animal for pleasure, see harmony in nature, and lend a helping hand to all living creatures,' say the Hindu Scriptures.

"Whenever we kill any animal for our food or pleasure we are selfish. It is on account of extreme selfishness that we do not recognise the rights of other animals, and that we try to nourish, nay, even to amuse ourselves, by killing innocent creatures or by injuring them, or by depriving them of their rights. This kind of selfishness is the mother of all evil thoughts and wicked deeds. That which makes us selfish and helps us to cling to our lower self is degrading and wicked ; that which leads us towards unselfishness is elevating and virtu-ous. That which prevents us from realising the oneness of Spirit is wrong ; that which opens our spiritual eyes and helps us to see that Divinity is expressing itself through the forms of lower animals, and makes us love them as we love our own Self, is godly and divine."

In the light of all the foregoing facts we can see that we have much to learn in our relations with the animal world from the Hindu people. They have grasped far more fully than we the great fact of the universe—namely, the essential unity, the essential one-ness of Life. When we have fully grasped this great fact, and when we live fully in accordance with it, then our civilisation will become a symmetrical civilisation, it will become all-round and complete, and not the one-sided and at times questionable civilisation it is at present. It will then be a blessing to all nations, to all the peoples of the earth, and not as it is so often to day in *some* respects, a verit-able curse and cause of degradation to them, for it will revolutionise in many

respects our relations and our dealings
with them.   It will also serve to make
perpetual that which, if we are not careful, may be merely transitory, just as it
has proved in the cases of many apparently strong and powerful nations before
us.   Let us follow the injunction of one
of the speakers at a meeting of the Society
for the Prevention of Cruelty to Animals
in a city of what some term a heathen
country, when, in urging an increase of
the Society's membership to at least 50,000,
he called upon the people to "write mercy
in the woods where the wild deer runs,
and in the air where our birds fly, and
all along the paths where our children and
our youths pass to and fro."

Our Humane Education Societies, our
societies for the Prevention of Cruelty to
Animals, as well as our few clinics and
hospitals and homes for animals, are receiving support from the best types of men
and women, but they need a still greater
and a far more universal support than
they are at present receiving.   Interest
along this line is growing, however, and I
think the time is rapidly coming when
men and women of means, in making be-

quests for the founding or the maintaining of institutions, will think of making them for the animal world as readily as they now think of making them for the human world. And still more, the wiser, the kindlier, and the more far-seeing among us, will give liberally to the support of every institution, every movement that has for its work humane, heart-training, so that there will be less need for last resorts, so that in coming time prevention will take the place of distress and suffering. It is simply a stirring of thought that is needed. Practically all cases of cruelty and ill-usage, and all careless treatment, arise through thoughtlessness, or have at least their beginnings in thoughtlessness.

We must learn to sympathise with the animals about us. We must realise that they love life just as we love it, that they suffer just as we suffer, that they are hurt by harshness and threats as we are hurt by them, that they are influenced by our thoughts as we are influenced by the thoughts of one another, that they love kindly treatment and that they appreciate it as we do ourselves, that they love and form attachments just as we do.

THE ENDURING SOUL

It would be exceedingly interesting and valuable were there place in a little volume of this nature to relate numbers of incidents and stories in connection with the lives of various animals—incidents and stories showing their devotion to those to whom they have become attached and whom they love, their intelligence, their powers of memory, their discernment and reason. ' Many is the time that an animal, perhaps the dog especially, has thrown itself into danger to warn from danger or to save the life of a human being, owner, friend, or stranger, without any apparent thought of its own safety or life. It was but a few weeks ago that I noticed among the news items on the editorial page of my daily paper, that on a Swiss eminence a monument is to be erected to Bary, that splendid St Bernard dog who during his life saved the lives of some forty persons.

And there are other monuments that I know of, erected to commemorate the fidelity or the sacrificial service of animals. But how many thousands of monuments, bathed at times with grateful tears and

hallowed by loving memories, have taken
form in the minds and hearts of those into
whose lives various animals have come.
And when we look into its eyes and see
the soul of the animal look out upon us,
with all its love and its fear, its warmth of
feeling, its confidence, wherever possible,
as well as its strange questionings, is it
possible for us longer to remain among that
company who feel that there is a great gulf
fixed, eternally fixed, between man and the
animal, many of whom live far more consistent and honest lives than we at times
live ourselves. Personally I believe that
their endeavour to live true to their various
natures and to their highest, even if at
times they fall short of it as we do, is
something that will be just as enduring
in their lives as in ours, and that they are
destined to a continually higher life, the
same as each and every one of us. The
common Father of us all, of the animal
as of ourselves, caused no one of His
creatures to be brought into existence
in vain, or for a mere temporary time.
Where there is a soul, be it in animal or
in human form, it is destined to endure as
such, even though the form, the body with

which it is clothed upon, and through which
it manifests on any particular plane of existence, changes, and in time falls away, to
give place to a new type of body better
adapted to the environment into which it
goes.

In order to be as concrete as possible,
we have been considering concrete cases
of carelessness and abuse and torture to
the animal world from our hands.   But I
think we have seen sufficiently clearly
already that whenever and every time we
sin against or do violence to these, our
fellow-creatures, we ourselves, in some form
or another, reap of the kind that we sow.
This is inevitably and invariably true, and
there is no escape from it.   And so, instead
of being their arch-enemy, let the children,
above all, be taught to become friends to,
and to care for and protect, these, their
fellow-creatures.

Let them be taught to give them always
kind words, and kind thoughts as well.
Some animals are most sensitively organised.
They feel and are influenced by our
thoughts and our emotions far more generally than we realise, and in some cases even

more than many people are. And why should we not recognise and speak to the horse as we pass him in the same way as we do to a fellow *human* being? While he may not get our exact words, he nevertheless gets and is influenced by the nature of the thought that is behind, and that is the *spirit* of the words. Let them be taught to become friends in this way. Let them be taught, even though young, to raise the hand against all misuse, abuse, and cruelty. Let them be taught that the horse, for example, when tired, or when its load is heavy, needs encouragement just as a man or a woman needs it, and that the whip is not necessary, except, indeed, in cases where he has not been taught to respond to words, but only to the whip. The whip is now used ninety-nine cases out of a hundred where it is not only unnecessary, but entirely uncalled for.

An American traveller, when riding one day with Tolstoi, noticed that he never made use of a whip when driving, and remarked to him to that effect. " No," he replied, with a slight spirit of disdain, " I talk to my horses. I do not beat them." Let us be taught by and let us carry to

the children the example of this Christ-
like man.

Heart-Training

Were I an educator, I would endeavour to make my influence along the
lines of humane, heart-training my chief
service to my pupils. The rules and
principles and even facts that are taught
them will, nine-tenths of them at least,
by-and-by be forgotten, but by bringing
into their lives this higher influence, at
once the root and the flower of all that
is worthy of the name "education," I would
give them something that would place them
at once in the ranks of the noblest of the
race. I would give not only special attention and time to this humane education,
but I would introduce it into and cause it
to permeate all of my work. A teacher
with a little insight will be able to find
opportunities on every hand.

M. de Sailly, an eminent French teacher,
who for a number of years has been giving
systematic humane instruction in his school,
says :

"I have long been convinced that kindness to animals produces great results, and that it is not only a powerful cause of material prosperity, but also the *beginning of moral prosperity*. My manner of teaching it does not disturb the routine of the school. Two days in the week all our lessons are conducted with reference to this subject. In the reading class I choose a book upon animals, and always give useful instruction and advice. My copies for writing are facts in natural history, and ideas of justice and kindness to animals. I prove that by not overworking them, and by keeping them in clean and roomy stables, feeding them well, and treating them kindly and gently, a greater profit and larger crops may be obtained. I also speak of birds and certain small animals which are very useful to farmers.

"The results are exceedingly satisfactory. The children are less disorderly, and more gentle and affectionate to each other. They feel more and more kindly to the animals and have ceased to rob nests and kill birds. They are touched by the suffering of animals, and the pain they feel when they see them cruelly used moves others to pity and compassion."

Mr George T. Angell, President of the

American Humane Education Society, has said :

"Standing before you as the advocate of the lower races, I declare what I believe cannot be gainsaid,—that just so soon and so far as we pour into all our schools the songs, the poems, and literature of mercy towards these lower creatures, just so soon and so far shall we reach the roots, not only of cruelty, but of crime. . . .

"A thousand cases of cruelty can be prevented by kind words and humane education for every one that can be prevented by prosecution."

And let us hear another sentence or two from another educator, a superintendent of schools in one of our New England States, —a sentence or two from an appeal to his fellows in connection with humane education :

"Fellow-teachers, let us make our teaching stronger and richer. Let us give our pupils something varied and inviting. Let us reach out more. Let us reach out for and take in humane education. Too much so-called teaching is unskilled labour. Too many of us are buried in our text-books— are mechanical hearers of lessons, are mere word-jugglers, fact-pedlers, and mind-stuffers. Let us put away all these things and *teach*. Let us put brains and heart into our work. Let us become character-builders. Such work will compel people to realise the grandly important truth that teaching is the profoundest science, the highest art, the noblest profession."

Then, were I a mother, I would infuse this same humane influence into all phases of the child's life and growth. Quietly and indirectly I would make all things speak to him in this language. I would put into his hands books such as "Black Beauty," "Beautiful Joe," and others of a kindred nature. I would form in my own village or part of the city, were there not one there already, a Band of Mercy, into which my own and neighbours' children would be called; and thus I would open up another little fountain of humanity for the healing of our troubled times.

We have recently been at war with another nation. There is to-day much unrest and uncertainty in connection with our foreign relations and policies. These matters, vital as they are, are of but small import compared with the questions and the conflict in connection with the social situation within our own borders that we shall be compelled squarely to face within the coming few years; the beginning of this time is indeed already at our very doors. The state of affairs referred to, as also its rapidly increasing proportions, is sufficiently well-known to all to make it unnecessary for more to be said in regard to it. Many who will have a hand in the solution and adjustment of these matters are now in our schools and on our streets, and we are educating them. We can educate them to patience, kindness, equity, and reason, or to hot - headedness, rashness, cruelty, and anarchy. And if these questions are not adjusted peaceably and through the influence of the former qualities, then they will be precipitated, through conflict and a terrific destruction of life and property, at the hands of those of the latter qualities.

We have now such agencies as will, in the hands of a small body of hot-headed, heartless men, burn half a city in a single night. Though one is a wealthy parent, his son may be the poor man and the anarchist. Though another parent is poor, his son may be the millionaire, and one of such a type as to be

F

hated by the great toiling classes. Time has a strange method of changing conditions. Both need to be humanely educated, the one equally with the other; and upon how thoroughly they are so educated will depend the orderly adjustment and peaceable solution of this rapidly coming time.

One of the most beautiful and valuable features of the kindergarten education, which comes nearer the true education than any we have yet seen, is the constantly recurring lesson of love, sympathy, kindness, and care for the animal world. All fellowships thus fostered, and the humane sentiments thus inculcated, will return to soften and enrich the child's, and later the man's or the woman's life, a thousand or a million fold; for we must always bear in mind that every kindness shown, every service done, to either a fellow human being or a so-called dumb fellow-creature, does us more good than the one for whom or that for which we do it. The joy that comes from this open-hearted fellowship with all living creatures is something too precious and valuable to be given up when once experienced. To feel and to realise the essential oneness of all life is a steep, up which the world is now rapidly coming. Through it ethics is being broadened and deepened, and even religion is being enriched and vitalised. Many, in all parts of the world, whose thoughts and sympathies have reached this higher plane, are giving abundantly of their time to push forward this much-belated humane element in human life. Others are giving abundantly of their treasure, through which many thousands of humane publications are being circulated, homes for animals are being established, humane education is being fostered, and the work of the various humane organisations is being enlarged in its scope and possibilities.

The strongest and noblest types of men and women are never devoid of this tender, humane sympathy, which is ever quick to manifest itself in kindness and care for every living creature. There is a little incident in the life of Lincoln which I found a few days ago in a most valuable little book recently published, entitled "Songs of Happy Life":

"In the early pioneer days, when he was a practising attorney and 'rode the circuit,' as was the custom at that time, he made one of a party of horsemen, lawyers like himself, who were on their way one spring morning from one court town to another. Their course lay across the prairies and through the timber; and as they passed by a little grove where the birds were singing merrily, they noticed a little fledgling which had fallen from the nest and was fluttering by the roadside. After they had ridden a short distance, Mr Lincoln stopped and, wheeling his horse, said, 'Wait for me a moment. I will soon rejoin you'; and as the party halted and watched him they saw Mr Lincoln return to the place where the little bird lay helpless on the ground, saw him tenderly take it up and set it carefully on a limb near the nest. When he joined his companions one of them laughingly said, 'Why, Lincoln, what did you bother yourself and delay us for, with such a trifle as that?' The reply deserves to be remembered, and it is for this that I have told the story. 'My friend,' said Mr Lincoln, 'I can only say this, that I feel better for it.'"

Let us go from this to one other incident in his life. During that famous series of public debates in Illinois with Stephen A.

Douglas in 1858, Mr Douglas at one place said, " I care not whether slavery in the Territories be voted up or whether it be voted down, it makes not a particle of difference with me." Mr Lincoln, speaking from the fulness of his great sympathetic heart, replied with emotion : "I am sorry to perceive that my friend Judge Douglas is so constituted that he does not feel the lash the least bit when it is laid upon another man's back."

Such are the strong, the valiant, the royal men and women, those with this tender soul-pathos, loving, caring, feeling for, sympathising with, both their fellow human beings and their so-called dumb fellow-creatures ; recognising that we are all parts of the one great whole, all different forms of the manifestation of the Spirit of Infinite Life, Love, and Power that is back of all, working in and through all,—the life of all.

www.ingramcontent.com/pod-product-compliance
Lightning Source LLC
Chambersburg PA
CBHW020041030726
47499CB00007B/2533